I0833787

Artificial Intelligence and the Modern Marketer

PenThere.com is an independent publishing company committed to supporting new and established writers from all backgrounds. We publish a diverse range of works spanning various genres, including fiction, non-fiction, business, technology, and poetry.

We are proud to present this book to the world, as we believe it will make a meaningful contribution to the understanding and application of AI in the realm of marketing.

ISBN: 979-8-218-23383-9

First Edition: June 2023

This book is dedicated to my tribe. I'm blessed to have the family and loved ones that I do and I'm eternally grateful.

Table of Contents

Chapter 1: Introduction to AI and Marketing

Welcome to *Artificial Intelligence and the Modern Marketer*! This book serves as a concise and practical guide to understanding artificial intelligence (AI) and its impact on marketing concepts. The primary goal is to build a strong foundation of knowledge by evaluating current information, speculating about the future, and drawing comparisons between real-world AI and its portrayal in TV shows and movies.

The convergence of AI and marketing has revolutionized the marketing landscape, bringing forth a new era of possibilities and opportunities. AI has made a significant impact on how marketers approach their strategies and engage with consumers. To better understand this transformative relationship, we'll explore an analogy that captures the essence of AI and marketing collaboration.

In this analogy, AI takes on the role of an inexperienced yet powerful student, similar to the character Neo in *The Matrix*. Just as Neo possesses immense potential, so does AI technology, with its machine learning algorithms, natural language processing capabilities, and data analytics, hold incredible power to reshape marketing practices.

On the other hand, the market represents the wise teacher, Morpheus. Marketers, armed with their knowledge, experience, and understanding of consumer behavior, guide and mentor the AI student, unlocking its true potential. This partnership between AI and marketers brings together unique skillsets and perspectives, resulting in powerful marketing strategies.

The impact of AI on the marketing landscape is profound. By harnessing the capabilities of AI, marketers gain data-driven insights that inform decision-making, enable personalization, and optimize campaigns for maximum effectiveness. Just like Neo's journey of self-discovery, AI continually evolves and learns from the vast amounts of data it processes. It adapts to changing market dynamics, identifies patterns and trends, and provides valuable recommendations to marketers.

However, this collaboration between AI and marketers is not without challenges. Privacy concerns, ethical considerations, and balancing between automation and human creativity pose hurdles that marketers must navigate. Just as Neo faced obstacles in fulfilling his destiny, marketers need to overcome these challenges to fully embrace the potential of AI in their strategies.

In the chapters to come, we will delve deeper into the specific ways AI transforms marketing, such as AI-powered advertising, predictive marketing, and data-driven customer insights. Through these explorations, we will uncover how AI unlocks creativity, anticipates future trends, and creates meaningful connections with target audiences.

In this book, marketers will learn to unlock the true power of AI, just as Morpheus guided Neo to become "The One." Let's step into the realm where imagination meets data, and the collaboration between AI and marketers ushers in a new era of marketing excellence.

Chapter 2: AI-Powered Advertising: Unleashing Creativity and Personalization

Get ready to witness an incredible transformation of advertising through the power of artificial intelligence. In this chapter, we will explore how the world of AI powered advertising intersects with personalized marketing.

The Evolution of Advertising:

Let's rewind and think about the evolution of advertising. From the days of traditional print and broadcast media to the vast landscape of digital platforms the world of advertising has come a long way, but the main principle is the same. It boils down to finding where eyes are focused and placing messaging that influences a specific action. And though it sounds so simple, the diversity of content and entertainment options makes consumer attention a moving target for those of us looking to influence decision-making. The newspaper used to be the one-stop-shop for information; it had local news, world news, sports, entertainment, coupons, classifieds, comic strips, etc. Add in the daily home delivery and it made sense for advertisers to spend big money to have their messaging included. Now, consider the multitude of options the average person has to find news, entertainment, coupons, etc. There are so many choices that it's incredibly fragmented. Up until now, marketers have combated the fragmentation with diversified marketing strategies and a mix of managed service from vendors and programmatic buys. However, the need to stay up-to-date with marketing trends and analyze mountains of marketing data is increasing. This is a key reason why AI will almost certainly become an indispensable tool for marketers. The application of AI in advertising is revolutionizing consumer targeting, content creation, and campaign optimization, cnabling brands to conncct with their audiences in more meaningful ways than ever before.

AI-generated Visuals, Voices, and Scripts:

Who would have guessed that AI would be just as strong on the creative side as it is in the quantitative world? One of the fastest growth categories has been with voice and visuals where algorithms harness the power of data to create stunning videos, captivating voice-overs, and compelling scripts for advertising campaigns. AI algorithms analyze vast amounts of information, extracting patterns and insights to craft engaging content that resonates with viewers. This fusion of AI-generated content and human creativity is a testament to the symbiotic relationship between technology and human ingenuity. Remarkably, there are even programs like Waymark AI that can generate a brand-safe commercial, script, and voiceover in a matter of seconds. While this process is evolving and far from perfect, it serves as a great tool for simple campaigns and brainstorming sessions. Just imagine the agility a marketing department gains when they can generate commercial drafts on the spot with just a few keywords and their company's website. On the other hand, imagine a smaller marketing department using this ability to compete with larger competitors because of the speed of use. Who doesn't love an underdog story? Soon we will be hearing about smaller brands outpacing corporate behemoths on a shoestring budget because of ingenuity and AI assistance.

Leveraging AI for Personalized Advertising:

The digital world is built on options and convenience, and businesses are striving to keep up with the individual needs of their customers. Marketing will continue to move in the same direction. Campaigns are more likely to resonate with potential consumers when said advertisements are personalized to the desired clientele. There are some advertisers that strive to get as close to the feel of a one-on-one conversation with their target audience as possible. That leads us to another moment where

AI gets to shine: sorting large amounts of client data and segmenting into personalized experiences. This is where AI takes the spotlight in delivering tailor-made experiences to consumers. Through the analysis of vast troves of consumer data, AI algorithms enable marketers to understand individual preferences, predict behaviors, and craft highly targeted campaigns. Personalized recommendations, dynamic ad placements, and customized messaging are some of the ways AI empowers brands to connect with their audience on a deeper level, fostering lasting engagement.

Bringing the Magic of Movies to Advertising:

Let's continue to draw inspiration from the world of movies and see how AI enhances advertising campaigns. Just as movies captivate audiences with their narratives and visuals, AI-driven techniques have also been employed to create memorable and impactful advertising experiences. From AI-generated movie trailers that capture the essence of a story to interactive ad campaigns that transport viewers into immersive narratives, AI infuses the magic of movies into the advertising realm.

Delving deeper, let's discuss the remarkable role of AI language models, such as ChatGPT, in the marketing funnel. These powerful AI-driven tools have the capability to revolutionize various aspects of marketing and empower marketers with their advanced functionalities.

One of the key strengths of AI language models is their ability to generate marketing text for social campaigns and outreach. With their natural language processing capabilities, they can analyze the target audience, understand the brand's objectives, and generate persuasive and compelling content that resonates with potential customers. Whether it's crafting engaging social media posts, impacting email newsletters, or captivating blog articles, AI language models like ChatGPT can assist marketers in delivering effective messages to their audience. To help illustrate ways AI language models can be used by marketers, several prompts have been included in Chapter 5, with additional prompts in the appendix.

In addition to generating content, AI language models can also perform edits based on specific marketing needs. They can review and refine existing marketing materials, ensuring coherence, clarity, and consistency in messaging. By leveraging AI language models, marketers can streamline their editing process and enhance the quality and impact of their content.

Furthermore, AI language models can serve as a powerful tool for generating or refining innovative ideas. Marketers can provide prompts or specific topics to AI models like ChatGPT, which can generate creative ideas, innovative concepts, and fresh perspectives that can inspire marketing campaigns. This capability can be particularly valuable when brainstorming new product launches, ad campaigns, or content strategies.

AI language models can also play a vital role in assisting with website content. They can help marketers optimize website copy by analyzing user data, understanding search intent, and generating SEO-friendly content that improves organic traffic and engagement. By leveraging AI models, marketers can enhance their website's user experience and drive conversions.

Moreover, AI language models can enhance presentations by generating compelling slides, providing data-driven insights, and suggesting impactful visual elements. They can also assist marketers in delivering persuasive pitches, engaging sales presentations, or informative webinars. By utilizing AI language models, marketers can captivate their audience and leave them with a memorable product.

The variety of applications for AI language models in marketing continues to evolve. With their ability to understand and process near-infinite amounts of data, these models have the potential to revolutionize numerous marketing tasks and processes. From content generation to idea refinement, website optimization to presentation enhancement, AI language models like ChatGPT can provide marketers with invaluable support and open new avenues for creativity and efficiency.

We have glimpsed the extraordinary possibilities that AI brings to advertising in this chapter. With AI as a creative partner, marketers can unleash their imagination and craft personalized campaigns that resonate with their target audience. The marriage of AI and marketing opens a world of untapped potential, where brands can captivate consumers in ways never before imagined.

Chapter 3: Predictive Marketing: Unlocking the Future

In the fast-paced world of marketing, staying ahead of the competition requires more than just guesswork. That's where predictive marketing comes into play. By harnessing the power of data and leveraging AI-driven analytics, marketers can unlock the ability to predict future outcomes and make informed marketing decisions. In this chapter, we will explore the fascinating realm of predictive marketing and how it is reshaping the marketing landscape.

Harnessing the power of data is the first step in predictive marketing. Marketers today have access to treasure troves of data, ranging from customer demographics and purchase history to online behavior and social media interactions. By leveraging AI technologies, this data can be processed, analyzed, and transformed into valuable insights that drive marketing strategies. Predictive analytics algorithms, powered by AI, can identify patterns, correlations, and trends within the data, enabling marketers to make predictions about customer behavior and market dynamics.

Returning to our movie analogies, the predictive power of AI is a common theme in many great films about the near and distant future. *Terminator*, *I-Robot*, and *Avengers: Age of Ultron* all explore the dangers of an artificial intelligence program becoming so adept at future thought that it decides to eliminate potential barriers before they have caused a problem for the program. Fortunately, real-life AI tools are far from dangerous; however, they tirelessly analyze and anticipate customer behavior. By analyzing historical data and identifying patterns, AI algorithms can predict future actions and preferences of individual customers. For instance, a predictive model can identify customers who are likely to churn and allow marketers to proactively engage with personalized retention strategies. Likewise, AI can predict customer preferences enabling marketers to recommend personalized products or services that create a more tailored and engaging customer experience.

Moreover, predictive marketing empowers marketers to stay ahead of market trends. By analyzing external data sources, such as social media trends, industry reports, and competitor analysis, AI algorithms can detect emerging market trends and anticipate shifts in consumer behavior. This enables marketers to adapt their strategies, launch new campaigns, or introduce innovative products that align with the evolving needs and desires of their target audience. It is important to note that focus groups, surveys and marketers' intuitions remain valuable tools. Predictive marketing, however, provides an additional layer of data-driven insights to enhance decision-making.

Real-life examples serve as powerful illustrations for the impact of predictive marketing. One such success story is Netflix, the popular streaming platform. By analyzing user behavior, viewing patterns, and preferences, Netflix's recommendation engine leverages predictive algorithms to suggest personalized content to its users. This not only enhances the user experience but also drives customer engagement and retention. Another notable example is the e-commerce giant, Amazon. Through its AI-powered recommendation system, Amazon analyzes customer browsing and purchasing history to predict and recommend products that align with each customer's preferences. This personalized approach not only increases sales but also fosters customer loyalty and satisfaction.

As predictive marketing continues to evolve, it is crucial for marketers to embrace the power of AI and data-driven insights. By leveraging predictive analytics, marketers can unlock the future, anticipate customer behavior, and adapt their strategies to meet evolving market trends. The era of guesswork is being replaced by informed decision-making, powered by AI algorithms and data-driven intelligence.

Chapter 4: Enhancing Customer Experience with AI

AI plays a transformative role in customer experience management, extending far beyond personalized experiences. AI technologies enable businesses to analyze customer data, identify patterns, and gain insights into customer preferences and behavior. By understanding customers on a deeper level, businesses can anticipate their needs and proactively address their pain points, ultimately enhancing the overall customer experience. AI-powered customer experience management enables businesses to deliver seamless interactions and build long-term relationships with their customers.

AI-powered chatbots and virtual assistants have emerged as valuable tools in customer experience management, but they are still far from reaching the level of sophistication exhibited by the fictional character Data from *Star Trek: The Next Generation*. However, envisioning a future where AI assistance reaches Data's level of intelligence and capabilities sparks excitement about the possibilities that lie ahead.

These intelligent systems leverage natural language processing and machine learning algorithms to understand customer inquiries and provide prompt assistance. AI chatbots can handle a wide range of customer queries, from basic information requests to complex problem-solving. They can even offer product recommendations based on customer preferences and purchase history, creating personalized experiences. By utilizing AI chatbots, businesses can deliver efficient customer support, improve response times, and ensure consistent service across different channels. With that said, while the systems continue to evolve there will still be many advantages to not fully automating this process. It is very easy to envision a system that starts with a proficient AI system that can be escalated to human interaction as needed.

Personalization and hyper-targeted marketing have become essential strategies for businesses to engage customers effectively. AI algorithms can segment customers based on various factors and deliver personalized messages, offers, and recommendations to each segment. This hyper-targeted approach enhances customer engagement, increases the likelihood of conversions, and fosters long-term loyalty. By harnessing AI-powered personalization, businesses can create relevant and meaningful interactions with customers, strengthening their brand relationships.

Another area where AI excels is in creating seamless omni-channel experiences. Omni-channel retailing requires businesses to provide consistent experiences across various channels, both online and offline. AI-powered systems can track customer interactions, preferences, and purchase history across multiple touchpoints, allowing businesses to deliver personalized experiences at every stage of the customer journey. AI enables businesses to provide cohesive brand experiences, seamless transitions between channels, and relevant recommendations based on customer behavior. By leveraging AI and integrating data from different channels, businesses can ensure a smooth and connected omni-channel experience for their customers.

By incorporating AI into customer experience management, businesses can transform how they interact with their customers. From AI-powered chatbots to hyper-targeted marketing and seamless omni-channel experiences, AI enhances personalization, efficiency, and overall customer satisfaction. Therefore, businesses can unlock new possibilities and create exceptional experiences that leave a lasting impression on their customers.

Chapter 5: AI-Powered Content Creation

In this chapter, we will cover AI-powered content creation. As businesses strive to engage their audiences with compelling and relevant content, AI technologies, such as natural language generation (NLG) and automated copywriting, have emerged as powerful tools. We explore the benefits, challenges, and ethical considerations associated with using AI in content creation processes, including blog articles, social media posts, and product descriptions.

The Rise of AI in Content Creation:

AI technologies have revolutionized various aspects of content creation, enabling businesses to automate and enhance their writing processes. NLG algorithms analyze data, generate human-like text, and produce content on a massive scale.

Benefits of AI-Powered Content Creation:

AI-powered content creation offers numerous advantages for marketers. Here are some benefits of using AI to generate blog articles, social media posts, and product descriptions:

1. Enhanced Efficiency: AI can produce content at scale and speed, reducing the time and

effort required for manual writing tasks. This allows marketers to focus on higher-level strategies and creative aspects of content creation.

2. Improved Personalization: AI algorithms can analyze user data and preferences to generate personalized content tailored to specific audiences. This level of personalization enhances customer engagement and increases the likelihood of conversions.

3. Consistent Quality: AI-powered content creation ensures consistent quality, adhering to brand guidelines and maintaining a coherent voice across multiple channels. This consistency helps in building a strong brand identity and trust with the audience.

4. Increased Creativity: AI tools can inspire and support human writers by suggesting creative ideas, generating unique angles, and assisting in the ideation process. By combining the creative abilities of humans with the data-driven insights of AI, marketers can create more compelling and innovative content.

Challenges and Considerations:

While AI-powered content creation brings significant benefits, it also presents challenges and ethical considerations that need careful attention:

1. Authenticity and Tone: AI-generated content may lack the human touch and authenticity that resonates with audiences. Marketers must strike a balance between efficiency and maintaining a genuine connection with their readers. Ensuring that the content has a human touch and aligns with the brand's voice is crucial.

2. Ethical Use of AI: Marketers need to be aware of potential biases in AI algorithms and ensure that AI-generated content adheres to ethical standards, avoiding misleading or manipulative practices. It is essential to use AI responsibly and ensure transparency in content creation processes.

3. Protecting Brand Voice: It's crucial to ensure that AI-generated content aligns with the brand's tone, values, and messaging, maintaining consistency and building trust with the audience. Marketers should establish clear guidelines to ensure that the content reflects the brand's identity accurately.

4. Content Originality and Plagiarism: AI tools should be used responsibly to avoid unintentional plagiarism or the creation of content that infringes on intellectual property rights. Marketers must ensure that AI-generated content is original and does not violate any copyright laws.

Best Practices for AI-Driven Content Creation:

To leverage the power of AI in content creation effectively, a set of best practices has been provided for marketers:

1. Human-AI Collaboration: Foster a collaborative approach where AI tools augment human creativity and expertise rather than replace them, ensuring that the final content reflects the brand's unique perspective. Human input is invaluable in adding the necessary nuances and emotions to the content.

2. Quality Assurance: Implement robust review and editing processes to ensure AI-generated content meets the desired standards of quality, accuracy, and relevance. Human editors play a vital role in fine-tuning the content and ensuring it aligns with the overall content strategy.

3. Continuous Learning and Improvement: Regularly evaluate the performance of AI-generated content, gather feedback from the audience, and refine the algorithms to enhance the output over time. By continuously learning from the data and user feedback, marketers can improve the effectiveness of AI-powered content creation.

4. Ethical Guidelines and Oversight: Establish clear ethical guidelines for the use of AI in content creation, ensuring compliance with legal and industry standards while prioritizing transparency and user trust. This includes monitoring and addressing any potential biases in the AI algorithms.

AI-Powered Content Creation Examples:

It is important for marketers to explore the potential of AI language models by leveraging various prompts to spark creative ideas. Here are a few examples:

1. Prompt: "Provide me with ideas for creating engaging social media content." Sample Response: "To create engaging social media content, consider using eye-catching visuals, incorporating storytelling elements, asking questions to spark conversations, running contests or giveaways, sharing user-generated content, and leveraging trending hashtags to join relevant conversations."

2. Prompt: "Provide tips for improving customer retention and loyalty." Sample Response: "To improve customer retention and loyalty, focus on delivering exceptional customer experiences, personalized communication, and rewards. Offer loyalty programs, exclusive

discounts, or special perks for repeat customers. Engage with customers through personalized emails, social media interactions, and customer feedback surveys. Provide proactive customer support and address any issues promptly to build trust and loyalty."

These prompts and answers demonstrate the capabilities of AI language models in generating useful information for content creation. For more examples and inspiration, please refer to the appendix.

AI-powered content creation has the potential to revolutionize the way marketers create and distribute content. By harnessing the capabilities of AI technologies, businesses can achieve higher efficiency, personalized experiences, and consistent quality. However, it is essential to navigate the challenges and ethical considerations associated with AI-generated content, maintaining authenticity and upholding ethical standards. By embracing the collaboration between human creativity and AI-driven assistance, marketers can unlock the full potential of AI in content creation and deliver impactful experiences to their audiences.

Remember, AI is an ever-evolving field, and staying informed about emerging trends and technologies is crucial for long-term success.

Continue to seek out information and explore the possibilities of AI in content creation to stay ahead in the dynamic digital landscape.

Chapter 6: AI-Powered Visual Content Recognition

In the digital age, visual content has become a powerful tool for marketers to engage and communicate with their audience. The emergence of AI technologies has revolutionized the way visual content is analyzed and utilized in marketing strategies. In this chapter, we will explore the interesting world of AI-powered visual content recognition and its applications in various marketing domains. From image recognition to user-generated content moderation and visual search optimization, AI is transforming the way marketers leverage visual content for their campaigns.

1. Image Recognition: Visual content is abundant on the internet, and AI-powered image recognition algorithms have the ability to automatically identify and categorize images at scale. By utilizing deep learning and computer vision techniques, AI can analyze the visual elements within an image, detect objects, and provide valuable insights for marketers. For example, image recognition can be applied in content tagging, allowing marketers to categorize their visual content more efficiently. It can also be used in brand monitoring to identify instances where a brand's logo or product is featured in

user-generated content. Additionally, sentiment analysis can be conducted by analyzing the visual components of an image to gain insights into the emotions it conveys.

2. User-Generated Content Moderation & Maintaining Brand Integrity: With the rise of social media and online platforms, user-generated content (UGC) has become a significant component of marketing strategies. However, ensuring the quality and appropriateness of UGC can be a daunting task. AI-powered visual content recognition plays a vital role in automating the moderation process. By employing AI algorithms to analyze images and videos, marketers can efficiently detect and filter out content that violates guidelines or poses risks to brand integrity. However, it's important to note that AI still has limitations in accurately identifying certain types of content. Some content creators have found ways to manipulate visuals to trick AI moderations systems. Therefore, human intervention is still necessary to ensure accurate content moderation and to handle cases where AI might fail to identify problematic content.

3. Visual Search Optimization & Enhancing Discoverability and Conversion: Visual search is revolutionizing the way consumers discover and interact with products. By using AI algorithms,

visual search engines can understand the visual attributes of an image and match it with relevant products or information. Marketers can leverage visual search optimization to enhance discoverability, drive conversions, and provide a seamless user experience. For example, by optimizing visual content with appropriate metadata and tags, marketers can increase the chances of the products appearing in visual search results. This allows consumers to easily find products that match their preferences and increase the likelihood of conversion.

AI-powered visual content recognition is reshaping the marketing landscape, providing marketers with powerful tools to analyze, moderate, and optimize visual content. From image recognition to user-generated content moderation and visual search optimization, AI is enabling marketers to unlock new possibilities in engaging their audience and driving business growth. As the field of AI continues to evolve, it is crucial for marketers to stay informed about the latest advancements and leverage them to create impactful visual marketing campaigns.

Chapter 7: Ethical Considerations in AI Marketing

The ethical implications of AI in marketing are significant. While AI brings unprecedented capabilities to marketing, empowering businesses to analyze vast amounts of data, automate processes, and deliver personalized experiences, it also raises ethical concerns that must be addressed. Marketers need to navigate these ethical dilemmas to ensure that AI-driven marketing practices are transparent, fair, and respectful of consumer rights.

In the rapidly evolving landscape of AI, legislation and regulations struggle to keep pace with advancements. It is the responsibility of companies to set firm boundaries and define their ethical stance regarding the use of AI tools. This involves considering the impact of AI on individuals, society, and the overall well-being of consumers.

Transparency, privacy, and data protection are crucial ethical considerations in AI marketing. Marketers leverage data to understand customer preferences, tailor marketing messages, and deliver personalized experiences. However, data collection and usage must be conducted in a transparent and ethical manner. Respecting privacy rights, obtaining proper consent, and safeguarding sensitive data are essential. Marketers should proactively communicate their data practices to consumers, provide opt-out options, and adhere to privacy regulations to maintain trust and protect individual privacy.

In the movie *Minority Report*, the concept of a society driven by data and predictive algorithms raises questions about privacy and consent. Just as individuals in the movie were concerned about their personal data being used without their knowledge, marketers must prioritize transparency and respect for consumer rights in the use of AI-driven marketing practices.

Transparency plays a vital role in combating misinformation and alleviating concerns related to data privacy. The complexity of digital marketing can be overwhelming for consumers, leading to fear about the rise of digitized personal data. By openly sharing information about data collection and usage, marketers can help build trust and alleviate concerns.

Addressing bias and ensuring fairness is another crucial ethical consideration in AI marketing. AI systems learn from data, but if that data is biased, it can perpetuate and amplify existing biases. In the marketing context, bias can result in discriminatory targeting, exclusion, or unfair treatment of certain groups. Marketers need to be aware of potential biases in data and algorithms and actively work to mitigate them.

Similar to the infiltration of biased algorithms in *Captain America: The Winter Soldier*, marketers must be diligent in identifying and mitigating potential biases in their AI systems. It is crucial to ensure fairness and inclusivity by diversifying training data and incorporating ethical guidelines into algorithm design.

Regularly auditing and evaluating AI systems, diversifying training data, and incorporating ethical guidelines into algorithm design can help promote fairness and inclusivity in AI marketing practices. Marketers must also stay informed about changing federal and state laws to ensure compliance with discrimination laws. Human oversight and accountability in AI-driven marketing are necessary to avoid blaming AI tools for unlawful marketing efforts.

Building trust through responsible AI practices is paramount in AI marketing. Trust is the foundation of any successful marketing strategy; AI marketing is no exception. Marketers must earn and maintain the trust of their audience by employing responsible AI practices. This includes being transparent about the use of AI, providing clear information about data collection and usage, and giving consumers control over their data.

Additionally, marketers should regularly evaluate the impact of AI on their marketing strategies, monitor for unintended consequences, and make necessary adjustments to ensure ethical and responsible practices. By demonstrating a commitment to responsible AI use and addressing ethical concerns head-on, marketers can foster trust and strengthen their relationships with customers.

The ethical considerations in AI marketing are essential for long-term success. By navigating these concerns with transparency, fairness, and responsibility, marketers can build trust, maintain consumer confidence, and contribute to the development of an ethical and sustainable AI-driven marketing ecosystem.

It is important to note that the responsible use of AI is an ongoing process, requiring continuous evaluation and adaptation as technologies evolve and societal expectations change. Marketers must stay vigilant, engage in ongoing education, and collaborate with industry peers to establish best practices that prioritize ethical considerations in AI marketing.

Chapter 8: The Future of AI and Marketing

First of all, congratulations on reaching the final chapter of our journey through the world of AI and marketing. In this chapter, we will consider the exciting prospects that lie ahead, including emerging trends and technologies, the impact on job roles and skillsets, the importance of human-AI collaboration, and the opportunities and challenges that await us in the AI-driven future of marketing.

Emerging Trends and Technologies in AI Marketing:

The field of AI is rapidly evolving, constantly introducing new trends and technologies that reshape marketing strategies. Advancements in natural language processing and computer vision are revolutionizing how marketers analyze and understand customer data. AI-driven personalization and hyper-targeting are enabling businesses to deliver highly tailored experiences to their customers. By staying at the forefront of these developments, marketers can gain deeper customer insights, deliver personalized experiences, and optimize their marketing campaigns for better results. Embracing these technologies will be crucial for businesses to stay competitive in the ever-evolving digital landscape. It also means that modern marketers must dedicate themselves to being life-long learners, as standard practices become outdated quickly. This book's primary goal is to help establish a foundation of knowledge that a savvy marketer will build upon throughout a long and successful career.

AI's Impact on Job Roles and Skillsets:

The rise of AI undoubtedly has implications for job roles and skill requirements within the marketing industry. While some may fear that AI will replace human marketers, the reality is a more symbiotic relationship. AI technologies can automate repetitive tasks, analyze vast amounts of data, and uncover valuable insights, allowing marketers to focus on higher-level strategic thinking and creativity. Marketers will need to adapt and acquire new skills that complement AI, such as data analysis, interpreting AI-generated insights, and creative problem-solving. By embracing AI as a tool, marketers can augment their capabilities and enhance their overall effectiveness in driving marketing success.

Human-AI Collaboration in Marketing:

The future of marketing lies in the collaboration between humans and AI. Just as extraordinary characters like Neo from *The Matrix* rely on the guidance of wise mentors like Morpheus, marketers and AI can form a powerful alliance. AI brings data-driven insights, automation, and personalization capabilities, while human marketers contribute creativity, intuition, and empathy. This collaboration can result in more impactful marketing strategies, where AI assists in data analysis and campaign optimization, while human marketers provide the strategic vision and personal touch. By fostering this collaboration, businesses can leverage the strengths of both humans and AI, creating marketing initiatives that resonate with customers on a deeper level.

Embracing the Opportunities and Challenges of the AI-driven Future:

The AI-driven future holds immense opportunities for marketers, but it also presents challenges that must be navigated carefully. As AI becomes more integral to marketing strategies, ethical considerations, privacy concerns, and the potential for bias come to the forefront. Marketers must embrace AI while ensuring transparency, fairness, and responsible use of data. By prioritizing ethical practices and adhering to privacy regulations, businesses can build trust with their customers and mitigate potential risks. It is essential to strike a balance between the power of AI and the human-centric approach to marketing, ensuring that AI serves as a tool to enhance customer experiences and drive business growth.

Embracing the AI-Driven Marketing Revolution:

In conclusion, we reflect on the key insights gathered throughout our exploration of AI and marketing. We emphasize the importance of embracing AI as a transformative force in marketing strategies. The future is ripe with possibilities, and those who adapt and integrate AI into their marketing efforts will be well-positioned for success. By staying curious, embracing new technologies, and cultivating a culture of continuous learning, marketers can harness the power of AI to unlock exceptional value for their customers and achieve remarkable business outcomes in this dynamic and ever-changing digital age.

As we end our journey, we reveal a fascinating revelation: this book, *Artificial Intelligence and the Modern Marketer* had a round of edits that was conducted by an AI language model. Additionally, the language model assisted with a portion of the brainstorming process. Just as we have emphasized the importance of collaboration and working together, we demonstrate the potential of AI to contribute to the creation of knowledge and insights in the field of marketing. I encourage readers to continue exploring the possibilities of AI in marketing and to seek further information or insights on the process. If you are curious about the details of collaboration, I encourage you to reach out directly.

The possibilities are boundless, from personalized experiences to predictive analytics, from hyper-targeted campaigns to seamless customer journeys. By embracing AI, marketers can unlock the potential to deliver exceptional value to their customers and achieve remarkable business outcomes in this dynamic and ever-changing digital age. The AI-driven marketing revolution awaits, and it is up to marketers to seize the opportunities and shape the future of marketing with the power of AI.

References:

Acquisti, A., Brandimarte, L., & Loewenstein, G. (2015). Privacy and human behavior in the age of information. Science, 347(6221), 509-514.

Bell, R. M., & Koren, Y. (2007). Lessons from the Netflix prize challenge. ACM SIGKDD Explorations Newsletter, 9(2), 75-79.

Breiman, L. (2001). Statistical modeling: The two cultures. Statistical Science, 16(3), 199-231.

Brynjolfsson, E., & McAfee, A. (2017). The business of artificial intelligence. Harvard Business Review, 95(1), 59-68.

Captain America: The Winter Soldier [Film]. (2014). Directed by A. Russo & J. Russo. Marvel Studios.

Davenport, T. H., & Harris, J. G. (2017). Competing on analytics: The new science of winning. Harvard Business Press.

Davenport, T. H., & Ronanki, R. (2018). Artificial intelligence for the real world. Harvard Business Review, 96(1), 108-116.

Desai, P., & Nair, S. (2017). Machine learning algorithms: A review. International Journal of Computer Science and Information Technologies, 8(2), 895-900.

Dhar, V. (2020). Artificial intelligence and the future of marketing. Journal of Marketing Analytics, 8(2), 87-90.

Goh, J., & Nachum, S. (2017). A framework for artificial intelligence capabilities in firms: What makes an organization AI-ready? Academy of Management Perspectives, 31(4), 266-278.

Johnson, J., & Singh, A. (2019). Driving customer engagement through predictive analytics: An empirical analysis of social media and purchase behavior. Journal of Marketing Analytics, 7(4), 205-219.

Kumar, V., & Raju, S. (2020). Artificial intelligence and marketing: The transformative power of machine learning and deep learning. Journal of the Academy of Marketing Science, 48(3), 413-414.

Li, H., Suomi, R., & Yoo, Y. (2009). Analyzing the impact of management practices on operational performance: A case study in manufacturing industry. International Journal of Services and Operations Management, 5(3), 331-348.

Marvel Studios. (2015). *Avengers: Age of Ultron* [Film]. Walt Disney Studios Motion Pictures.

The Matrix [Film]. (1999). Directed by L. Wachowski & A. Wachowski. Warner Bros. Pictures.

Minority Report [Film]. (2002). Directed by S. Spielberg. DreamWorks Pictures & Twentieth Century Fox.

O'Leary, D. E. (2012). How to measure and manage your company's reputation. MIT Sloan Management Review, 53(3), 37-44.

Provost, F., & Fawcett, T. (2013). Data science and its relationship to big data and data-driven decision making. Big Data, 1(1), 51-59.

Star Trek: The Next Generation [Television series]. (1987-1994). Created by G. Roddenberry. Paramount Television.

Terminator [Film]. (1984). Directed by J. Cameron. Hemdale Film Corporation & Pacific Western Productions.

Waymark AI. (n.d.). Retrieved from Waymark.com.

Appendix

Sample ChatGPT Prompts for Marketers:

Here are 10 quick prompts to help jumpstart your brainstorming process:

1. Prompt: "Provide me with ideas for creating engaging social media content."

Sample Response: "To create engaging social media content, consider using eye-catching visuals, incorporating storytelling elements, asking questions to spark conversations, running contests or giveaways, sharing user-generated content, and leveraging trending hashtags to join relevant conversations."

2. Prompt: "Help me brainstorm creative ways to leverage AI in my marketing campaigns."

Sample Response: "AI can be leveraged in marketing campaigns by utilizing chatbots for customer support, personalizing content recommendations based on user behavior, optimizing ad targeting through predictive analytics, automating email marketing workflows, and implementing AI-powered chat or voice assistants for interactive experiences."

3. Prompt: "Suggest strategies to improve my email marketing open rates and click-through rates."

Sample Response: "To improve email open rates and click-through rates, consider optimizing subject lines to be concise and compelling, personalizing email content based on recipient data, segmenting your email list for targeted campaigns, optimizing email design for mobile devices, including clear and prominent call-to-action buttons, and conducting A/B testing to identify the most effective elements."

4. Prompt: "How can I optimize my website for better search engine rankings?"

Sample Response: "To optimize your website for better search engine rankings, focus on keyword research and incorporate relevant keywords into your website content, meta tags, and URLs. Improve website loading speed, ensure mobile responsiveness, create high-quality and shareable content, build backlinks from reputable sources, and regularly update your website with fresh content."

5. Prompt: "Assist me in creating a customer persona for my target audience."

Sample Response: "To create a customer persona, start by researching your target audience demographics, interests, and behavior. Identify their pain points, motivations, and goals. Develop a detailed profile that includes their age, gender, occupation, hobbies, preferences, and challenges. This persona will help you tailor your marketing strategies to better resonate with your target audience."

6. Prompt: "Provide tips for improving customer retention and loyalty."

Sample Response: "To improve customer retention and loyalty, focus on delivering exceptional customer experiences, personalized communication, and rewards. Offer loyalty programs, exclusive discounts, or special perks for repeat customers. Engage with customers through personalized emails, social media interactions, and customer feedback surveys. Provide proactive customer support and address any issues promptly to build trust and loyalty."

7. Prompt: "How can I leverage influencer marketing to promote my brand?"

Sample Response: "To leverage influencer marketing effectively, start by identifying influencers who align with your brand values and target audience. Reach out to them with a compelling pitch, offering mutually beneficial partnerships. Collaborate with influencers on content creation, such as sponsored posts, product reviews, or takeover campaigns. Encourage them to share their authentic experiences with your brand to generate buzz and reach their dedicated followers."

8. Prompt: "Suggest strategies for improving conversion rates on my e-commerce website."

Sample Response: "To improve conversion rates on your e-commerce website, ensure a user-friendly and intuitive website design with clear navigation and prominent calls-to-action. Optimize product descriptions and images to provide accurate and compelling information. Simplify the checkout process, offer secure payment options, and implement social proof elements such as customer reviews and testimonials. Use scarcity tactics like limited-time offers or low stock notifications to create a sense of urgency."

9. Prompt: "How can I effectively measure the ROI of my marketing campaigns?"

Sample Response: "To measure the ROI of your marketing campaigns, set clear goals and key performance indicators (KPIs) aligned with your objectives. Use tracking tools like Google Analytics to monitor website traffic, conversions, and customer behavior. Implement unique campaign URLs and trackable links to measure the impact of specific marketing channels. Calculate the cost per acquisition (CPA), return on ad spend (ROAS), and customer lifetime value (CLTV) to assess the effectiveness of your campaigns."

10. Prompt: "Help me develop a content marketing strategy for my business."

Sample Response: "To develop a content marketing strategy, start by defining your target audience and their informational needs. Create a content calendar outlining topics, formats, and publishing frequency. Produce high-quality, valuable content such as blog posts, videos, infographics, or podcasts. Promote your content through various channels, including social media, email newsletters, and guest posting. Measure engagement metrics and feedback to refine your strategy over time."

Remember, these are just sample responses, and the actual answers generated by ChatGPT may vary. It's important to review and refine the responses according to your specific needs and industry context.

Prompts For Idea Expansion

There will be times where you'll want your AI Language Model to expand on an idea. Here are a few examples of useful prompts and sample answers:

1. Prompt: "Explain the importance of social media marketing in today's digital landscape and provide tips for maximizing its impact."

   Answer: Social media marketing plays a pivotal role in today's digital landscape. It allows businesses to connect with their target audience, build brand awareness, and drive engagement. To maximize its impact, marketers should:

   - Identify the most relevant social media platforms for their target audience.
   - Develop a consistent content strategy that aligns with the brand's voice and values.
   - Leverage social media analytics to measure performance and optimize campaigns.
   - Encourage audience interaction and participation through contests, polls, and user-generated content.
   - Collaborate with influencers to amplify brand reach and credibility.

2. Prompt: "Discuss the role of customer segmentation in developing targeted marketing campaigns and share best practices for effective segmentation."

   Answer: Customer segmentation is crucial for developing targeted marketing campaigns. By dividing the target audience into distinct segments based on demographics, behaviors, or preferences, marketers can deliver personalized messages that resonate with each group. Best practices for effective segmentation include:

   - Conducting thorough market research to identify relevant segmentation variables.
   - Using data analytics and customer insights to create accurate and meaningful segments.
   - Tailoring marketing messages and offers to address the specific needs and interests of each segment.
   - Continuously monitoring and updating segmentation strategies based on changing market dynamics.

3. Prompt: "Explore the benefits and challenges of incorporating video marketing into a marketing strategy, along with tips for creating compelling video content."

   Answer: Video marketing has become a powerful tool for capturing audience attention and conveying compelling brand messages. To create impactful video content and maximize its impact, consider the following tips:

   - Start with a strong hook to grab viewers' attention within the first few seconds.
   - Tell a story that resonates with your target audience, focusing on emotions and relatable experiences.
   - Keep videos concise and visually engaging, leveraging visuals, graphics, and animation.
   - Optimize videos for different platforms and devices to ensure seamless viewing experiences.
   - Encourage social sharing and engagement by incorporating calls to action and interactive elements.

4. Prompt: "Explain the concept of brand storytelling and its significance in building brand identity. Share examples of successful brand storytelling campaigns."

   Answer: Brand storytelling is a powerful way to build brand identity and connect with consumers on a deeper level. Successful brand storytelling campaigns often involve:

   - Identifying a brand's unique narrative, values, and mission.
   - Creating compelling and authentic stories that align with the brand's identity.
   - Leveraging various storytelling mediums, such as videos, social media posts, and blog articles.
   - Engaging with the audience through emotional storytelling, humor, or thought-provoking narratives.
   - Evoking a sense of community and inviting audience participation in the brand's story.

5. Prompt: "Discuss the impact of user-generated content (UGC) on brand credibility and engagement. Provide strategies for encouraging and leveraging UGC in marketing efforts."

   Answer: User-generated content (UGC) has a significant impact on brand credibility and engagement. To encourage and leverage UGC effectively, consider the following strategies:

   - Encourage customers to share their experiences and opinions through reviews, testimonials, or social media posts.
   - Showcase UGC on brand websites, social media channels, or marketing materials to build trust and authenticity.
   - Run contests or campaigns that encourage users to generate content related to the brand.
   - Engage with UGC by responding to user comments, acknowledging their contributions, and fostering a sense of community.
   - Obtain proper permissions and rights before using UGC in marketing materials to respect intellectual property.

6. Prompt: "Explain the concept of marketing automation and its benefits for streamlining marketing processes. Share examples of effective marketing automation strategies."

   Answer: Marketing automation streamlines marketing processes, allowing businesses to automate repetitive tasks and deliver personalized experiences at scale. Effective marketing automation strategies involve:

   - Mapping out customer journeys and identifying touchpoints for automation.
   - Implementing automated email campaigns triggered by specific user actions or behaviors.
   - Utilizing marketing automation tools to segment audiences, personalize content, and track campaign performance.
   - Integrating marketing automation with customer relationship management (CRM) systems for seamless data synchronization.
   - Continuously monitoring and optimizing automated workflows to ensure relevancy and effectiveness.

7. Prompt: "Discuss the evolving role of chatbots in customer support and lead generation. Highlight the benefits and considerations when implementing chatbot technology."

   Answer: Chatbots are revolutionizing customer support and lead generation. When implementing chatbot technology, consider the following benefits and considerations:

   - Chatbots provide instant support and assistance, improving response times and customer satisfaction.
   - They can handle common inquiries, freeing up human resources for more complex tasks.
   - Proper training and testing are essential to ensure accurate and helpful chatbot responses.
   - Maintain a balance between automated responses and human interaction to provide a personalized experience.
   - Regularly analyze chatbot interactions to identify areas for improvement and fine-tune responses.

8. Prompt: "Explore the concept of omnichannel marketing and its role in delivering consistent customer experiences across multiple channels. Share strategies for implementing successful omnichannel marketing campaigns."

   Answer: Omnichannel marketing ensures consistent customer experiences across multiple channels and touchpoints. Successful implementation involves:

   - Integrating customer data and insights across channels to provide a seamless journey.
   - Delivering consistent messaging, branding, and user experiences across all channels.
   - Personalizing content and offers based on user preferences and behaviors across different channels.
   - Providing flexible and convenient options for customers to interact and make purchases.
   - Leveraging technology, such as customer data platforms and marketing automation, to orchestrate omnichannel campaigns.

9. Prompt: "Discuss the growing importance of voice search optimization in SEO and provide tips for optimizing content for voice search."

   Answer: Voice search optimization is increasingly important in SEO as more users utilize voice assistants. To optimize content for voice search, consider these tips:

   - Target long-tail keywords and natural language queries that align with voice search patterns.

   - Optimize website content for featured snippets, as they often appear in voice search results.

   - Structure content with clear headings and concise answers to address specific voice queries.

   - Ensure website speed and mobile friendliness, as voice searches often occur on mobile devices.

   - Incorporate conversational language and provide direct, informative responses to voice queries.

10. Prompt: "Explain the concept of influencer marketing and its impact on brand awareness and consumer trust. Share strategies for identifying and collaborating with influencers in a specific industry or niche."

    Answer: Influencer marketing has a significant impact on brand awareness and consumer trust. When identifying and collaborating with influencers, consider the following strategies:

    - Research and identify influencers whose values and audience align with your brand.
    - Establish clear campaign objectives and expectations when collaborating with influencers.
    - Foster genuine relationships with influencers through mutual trust, open communication, and long-term partnerships.
    - Provide influencers with creative freedom to authentically promote your brand within their content.
    - Measure the success of influencer campaigns using key performance indicators such as reach, engagement, and conversions.

These sample answers provide insights and recommendations for each prompt, helping marketers understand and implement effective strategies in various areas of marketing.

These prompts can help you delve deeper into specific marketing topics and provide more comprehensive insights for each subject.

About the Author

Cortlan Booker is an accomplished digital marketing professional with over a decade of experience in implementing effective strategies for clients across various industries. He has a proven track record of delivering measurable results through a strategic approach. Cortlan earned his bachelor's and master's degree from Ball State University in Indiana and currently resides in Cincinnati, Ohio. Mr. Booker combines his deep understanding of business, technology, and relationship building to drive success for his clients.

Outside of his professional pursuits, Cortlan values the importance of personal connections and enjoys spending quality time with his loved ones. He finds inspiration in the world of sports, appreciating the teamwork, dedication, and drive for excellence that athletes exhibit. In his leisure time, Cortlan indulges in his love for music and movies, finding joy in the power of storytelling and its ability to captivate audiences.

www.ingramcontent.com/pod-product-compliance
Lightning Source LLC
LaVergne TN
LVHW010543100826
845148LV00013B/2575

* 9 7 9 8 2 1 8 2 3 3 8 3 9 *